John William Harshberger

The Vegetation of the Yellowstone Hot Springs

John William Harshberger

The Vegetation of the Yellowstone Hot Springs

ISBN/EAN: 9783743346598

Manufactured in Europe, USA, Canada, Australia, Japa

Cover: Foto ©berggeist007 / pixelio.de

Manufactured and distributed by brebook publishing software
(www.brebook.com)

John William Harshberger

The Vegetation of the Yellowstone Hot Springs

THE VEGETATION OF THE YELLOWSTONE HOT SPRINGS.

By John W. Harshberger, Ph.D.

The actual discovery of the Yellowstone Wonderland, by which is meant its full and final disclosure to the world, was the work of three parties, who visited and explored it in the years 1869, 1870 and 1871. Although, since the last date, much has been written concerning the geological and physiographical features of the park set aside by Act of Congress in the year 1872, little has been written concerning the flora of the region, and what has been published deals almost entirely with the plants from a systematic standpoint.

Situated in the northwestern corner of Wyoming, in the Rocky Mountains, at an elevation ranging from 6,000 to 12,000 feet, the region is one of high and lofty mountains, of deep cañons walled in by precipitous sides, and of beautiful upland valleys, the natural haunts of the timid herbivora that seek the mountain meadows for the tender and nutritious grasses which grow there luxuriantly. The pasturage in many of the meadows and valleys is excellent, being formed by the growth of such grasses as alpine timothy, *Phleum alpinum*, blue joint, *Calamagrostis Canadensis*, sheep's fescue, *Festuca ovina*, Kœleria, *Kœleria cristata*. The herbaceous vegetation is not so striking as in many other regions, but still the distribution of such species as do occur is interesting. In the lakes and rivers we find the aquatic vegetation to consist of *Ranunculus aquatilis*, *Nuphar advena*, *Nuphar polycephalum*, *Utricularia vulgaris*, *Lemna trisulca*, *Typha latifolia*, *Sparganium simplex*, etc. Near the head of Yellowstone Lake is found *Subularia aquatica*, a plant of quite a remarkable distribution, found nowhere else in America except in Maine and New Hampshire. *Gentiana detonsa*, *Spraguea umbellata* are striking plants. The meadows and hillsides are spangled with bright-colored flowers, among which may be noted the bee larkspur, *Delphinium Menziesii*, the columbine, *Aquilegia flavescens*, the harebell, *Campanula*, the aconite, *Aconitum Columbianum*, the lupine, *Lupinus*, the evening primrose, *Œnothera*, the aster, the painted cup, *Castilleia*. It is a remarkable fact that scarcely a night passes throughout the summer without frost, so that the herbaceous plants grow and bloom under somewhat unusual conditions. The fringed gentian, *Gentiana detonsa*, closes its flowers as night approaches, to open them again in the morning, and many other plants provided with a hairy or woolly covering are thus secure against frost action. The plants of the Yellowstone region, as far as observed, are well adapted to their surroundings.

The forests are formed by one tree predominating, *Pinus contorta*, *var. Murrayana*, which grows tall and straight, but never reaches any considerable girth. Interspersed among the pines we find several other arborescent species, namely, Douglass spruce, *Pseudotsuga Douglasii*, the largest tree in the park; balsam, *Abies subalpina*, pine, *Pinus Engelmannii*, red cedar, *Juniperus Virginiana*, poplar, *Populus tremuloides*, and willow, *Salix*, of several species. These forests are of great importance in conserving the rain which falls. Many of the most important rivers of the western United States rise in this region, the Missouri, the Yellowstone, the Wind, the Big Horn, the Platte, the Green (afterward the Colorado), and the Snake, which flows through Wyoming, Idaho and Washington, emptying into the Columbia, and thus reaches the Pacific.

Yellowstone Park, notwithstanding its wild grandeur as a mountain domain, is yet more interesting on account of the geological wonders which are found within its boundaries, namely the geysers and hot springs. The geysers are actively throwing up in jets at periodic intervals, steam and boiling water; the hot springs are either quiescent, or are bubbling and boiling without explosive eruption. They are found in four distinct areas in the Park; the geysers and the hot springs in the Upper, Lower and Norris Geyser Basin, hot springs only in the Mammoth Hot Spring Region. This division also accords with the predominating chemical content of the waters. In the Upper, Lower and Norris Geyser Basins, we have springs and geysers which are actively depositing silicious material (sinter); in the Mammoth Hot Spring Basin, springs which are forming calcareous deposits, called travertine.

Much inquiry has been instituted concerning the therapeutic value of the mineral springs of the Park. Many hot spring regions throughout Europe and America are resorted to by thousands in search of health. The hot springs of Virginia are visited by hundreds every year. It is said of the Yellowstone region, that the first explorers to ascend the Gardiner River, in 1871, found numbers of invalids encamped on the banks, where the hot waters from Mammoth Hot Springs enter the stream; and it is recorded that they were most emphatic in their favorable impressions in regard to their sanitary effects. No one now goes to the Park on account of its mineral waters. It would, therefore, be premature to assume that there is no medicinal virtue in them. Two great drawbacks are to be encountered, and these alone are sufficient to explain why the Yellowstone will probably never become a resort for invalids. Inaccessibility, length and severity of the winters are sufficient obstacles to the National Park ever becoming such a resort. The open summer season lasts only about three months.

The hot springs and geysers, on the other hand, are interesting to the geologist, because of the remarkable phenomena connected with their origin and activity; to the botanist they are fascinating, because of the low forms of vegetal life found existing in them even at high temperatures.

As before stated, the waters which run from the hot springs and geysers of the Yellowstone may be comprehended under two heads—those which deposit silica, as sinter, and those which form calcium carbonate, as travertine. The last-mentioned substance is only found in the Mammoth Hot Spring Basin; the latter makes up the characteristic formations of the Norris, Lower and Upper Geyser Basins. The question naturally arises, how are the beautiful terraces which surround many of the hot spring centres formed? Are they not simply built up by the deposition of new material from the overflow water, as it evaporates and cools at the surface? At first sight, it would seem that the craters and bowls of the geysers and hot springs were formed in this way, because we know that boiling water, under pressure, will dissolve and hold in solution much more inorganic material than ordinary river or spring water at the normal temperature, and that in many instances, when the pressure is relieved and the temperature lowered, the water will precipitate its mineral contents.

In the case of the richly carbonated waters of the Mammoth Hot Springs, calcium carbonate is deposited by the relief of pressure, by the escape of the carbon dioxide and by the evaporation of the water; but this physical process is not the sole cause of the varied and beautiful terraces, which will presently be described. At the Norris Geyser Basin, relief of pressure and cooling will cause a separation of silica from the hot waters, but the waters of the other geyser basins contain very much less silica, and, as far as has been observed by geologists, neither relief of pressure nor cooling will produce a separation of the silica. Water collected from the springs and geysers of the Upper and Lower Geyser Basins was perfectly transparent, and remained clear and without sediment after standing for several years. Experiments showed that the silica in these waters remained dissolved, even when the water was cooled down to the freezing point, and it was only after the crystallization of the water by freezing that the silica was separated and settled down as an insoluble flocculent precipitate upon melting the ice.

How, then, are we to account for the production of the exquisite terraces, mounds, pools and geyser cones? It has been proved, in addition to the causes operative in the above instance, that the rapid deposition of the sinter and travertine from both classes of water is due to the action of vegetation in removing the carbon dioxide from carbonated waters, thus depositing calcium carbonate, and, in the case of the silicious waters, depositing by the activity of the protoplasm a gelatinous silica, which, upon exposure, finally hardens. We know, from numerous observations, that plants are active in rock building and disintegration.

The plants of the Carboniferous Period, by their death and consolidation, formed the extensive and useful coal beds. Sphagnum and mosses compacted yield peat, and, in some cases, soft coal. Silicious diatoms have given rise to extensive diatomaceous earths. In several of the higher algæ, for example, *Halimeda opuntiæ*, the carbonate of lime deposited by the plant forms a sieve-like cover about the tips of the algal filaments, and, in *Acetabularia*, it occurs as a tube about the stalk of the plant. In the CHARÆ the lime is separated and deposited in the cells and cell walls of the back alone, while in the *Corallines* it is found only within the cells. Nor is our knowledge of the activity of protoplasm in the deposit of mineral substance solely confined to plants. We know that many animals secrete silex and carbonate of lime, foraminifera, coral polyps and molluscs generally. Before, however, we can understand the part which vegetation has played in forming the travertine and sinter beds of the Yellowstone Park, we must become familiar with the general appearance and character of the deposits themselves.

First in importance among the many points of interest accessible are the Hot Spring Terraces. These have been built one upon another, until the present active portion constitutes a hill rising 300 feet above the site of the Mammoth Hot Springs Hotel. The formation about these springs, it will be remembered, is calcareous, and to this fact is due its distinctive character, so different from the silica formations which prevail elsewhere in the Park. " The over-hanging bowls which these deposits build up are among the finest specimens of Nature's work in the world, while the water that fills them is of that peculiar beauty to be found only in thermal springs." Cleopatra Spring, Jupiter Terrace, Pulpit Terrace, Minerva Terrace, are among the most interesting and beautiful of the active springs. One of the most beautiful is a pool filled with pellucid water in vio-lent ebullition. The sides and bottom of the basin are formed of pure white travertine, while the varying depths cause the water to appear all shades of blue and green, from a deep peacock blue in the deeper parts of the bowl, to the lightest of Nile greens in the shallow re-cesses. In wandering about the terraces, one is much impressed with the brightly tinted basins about the springs, and the red and orange colors of the slopes overflowed by the hot waters. These colors are due to the presence of the microscopic plants, algæ of several forms and species. In the cooler springs and channels simi-lar vegetation forms the bright green, orange or brown mem-brane-like sheets, or masses of jelly without apparent vegetal struc-ture. Silken yellow filaments are found in bowls and channels of the hottest springs. Words fail to convey an adequate idea of the massive marble-like terraces, rising tier upon tier, and the exquisite coloring of their sides and the margins of the bowls filled with steaming hot water of most magnificent iridescent hues.

The silicious formations are similar, although not raised in ter-races so grand or imposing, simply because the formation of silicious sinter is much slower than the formation of the travertine, and because the region seems to be of later geologic age. Many of the geyser cones are bee-hive in shape, of a white adamantine-like appearance, and are, as a rule, delicately colored by pale greens and pinks of exquisite variation. The many hundreds of springs of the Upper Geyser Basin, where they are seen at their best, are generally characterized by the transparent clearness of the water, which appears of varying shades of blue and green, according to the depth and amount of light admitted. Morning Glory Spring is one of the most beautiful springs of the Park, with a funnel-sha ed cone sug-gesting the flower, and with walls most delicately colored.

Black Sand Basin is, however, most interesting for our purpose. The description of Dr. Peale is interestingly comprehensive, and is as follows: "This is one of the most beautiful springs in the Upper Basin. It has a delicate rim, with toadstool-like masses around it. The basin slopes rather gently toward a central aperture, that, to the eye, appears to have no bottom. The water in the spring has a delicate turquoise tint, and as the breeze sweeps across its surface, dispelling the steam, the effect of the ripple of the water is very beautiful. The sloping sides are covered with a light brown crust; sometimes it is rather a cream color. The funnel is about 40 feet in diameter, while the entire space covered by the spring is about 55 x 60 feet, outside the rim of which is a border of pitch stone (obsidian) sand or gravel, sloping 25 feet. From its west side flows a considerable stream, forming a most beautiful channel, in which the coloring presents a remarkable variety of shades; the extremely delicate pinks are mingled with equally delicate tints of saffron and yellow, and here and there shades of green."

The overflow from this spring spreads out over a large area, called Specimen Lake, where absorption of the silica from the water has destroyed many of the trees of the vicinity, the dry, lifeless trunks adding to the attractiveness of the place by affording the appearance of petrifactions.[1] All of these exquisite masses of colors which are found lining the pools, filling the overflow channels and spreading out flat in the lower marshy places, are due to the growth of vegetal organisms belonging to the bacteria and algæ.

Walter H. Weed[2] describes the appearance of the Black Sand Basin and channels filled with algal growths: "As the water from this spring flows along its channel it is rapidly chilled by contact with the air and by evaporation, and is soon cool enough to permit the growth of the more rudimentary forms which live at the highest temperature. These appear first in skeins of delicate white filaments which gradually change to pale flesh-pink farther down stream. As the water becomes cooler, this pink becomes deeper, and a bright orange and closely adherent fuzzy growth, rarely filamentous, appears at the border of the stream, and finally replaces the first-mentioned forms. This merges into yellowish-green, which shades into a rich emerald farther down, this being the common color of fresh-water algæ. In the quiet waters of the pools fed by this stream

[1] Haynes-Guptill, Guide to Yellowstone Park, p. 68.
[2] Weed, Ninth Annual Report U. S. Geological Survey, p. 657.

the algæ present a different development, forming leathery sheets of tough gelatinous material, with coralloid and vase-shaped forms rising to the surface, and often filling up a large part of the pool. Sheets of brown or green, kelpy or leathery, also line the basins of warm springs whose temperature does not exceed 140° F., but in springs having a higher temperature the only vegetation present forms a velvety, golden-yellow fuzz upon the bottom and sides of the bowl. This growth is rarely noticed in springs where the water exceeds 160° F., except at the edge of the pool. If the basin is funnel-shaped, with flaring or saucer-shaped expansion, algæ grow in the cooler and shallower water of the margin, forming concentric rings of yellow, old gold and orange, shading into salmon-red and crimson, and this to brown at the border of the spring. Around such springs the growth at the margin often forms a raised rim of spongy, stiff jelly, sometimes almost rubber-like in consistency, and red or brown in color. Evaporation of the water drawn up to the top of such rims leaves a thin film of silica, which thickens to a crust and so aids in the production of a permanent sinter rim."

Near some springs, for example near the Emerald Pool, algal channels are formed and the waterway is floored with a sheet of olive or emerald green, kelpy jelly. Where there is a moderate current, this lining is nearly smooth, resembling a sheet of wet leather, but in quieter waters this soft carpet is dotted with little warty excrescences, and little pillars produced by the upward growth of the algæ; the pillars sometimes terminate by balloon-like caps or globes containing bubbles of gas. When, by their upward growth, these pillars reach the surface of the pool, they increase rapidly in diameter, and form flat, cap-shaped formations which sometimes merge into table-like expansions of quite peculiar form. The continued growth of new pillars dams up the outlet, and the water collecting forms shallow lagoons or pools of varying degrees of temperature. As the temperature changes, the nature of the growth changes, the bright-colored algous jelly forming the outer covering of the pillars changes to light salmon-pink, and the substance itself becomes noticeably silicious, or forms a filmy web upon the silicious centre.

It has been for some time known that the hot springs of the world support various growths of microscopic plants. Agardh and Corda recognized and described such in the hot springs of Carlsbad,

Bohemia. Later, Cohn, in 1862, showed that the alg.æ of these springs deposited travertine. Sir William Hooker, in 1809, found CONFERVACE.E at the borders of many of the hot springs there. *Conferva limosa, C. flavescens, C. rivularis* were abundant in the water. Baring Gould, who visited the Icelandic geyser region in 1864, found in the overflow channels of the spring, Tunguhver, a species of the genus *Hyphcothrix*, common in hot waters all over the world. In New Zealand, the presence of algæ in hot springs has been determined. In the hot springs of the Azores, Mr. Moseley found algæ forming a pale yellowish-green layer an inch and a half thick. The temperature of the water was 176° F. to 194° F. A thick, brilliant green growth, consisting of *Chroococcus* was found at the edge of a shallow pool of hot water, where the temperature was between 149° F. and 156° F.

In the hot springs of the Yellowstone no plant life has been found at a temperature exceeding 185° F., some degrees below the boiling point of water, which, at the altitude of the park (7,000–8,500 feet) is 198° F. The most luxuriant growth of algæ is found in water which has cooled down to a temperature of 104° F. to 122° F. In water of a temperature ranging from 100° F. to 125° F., we have the greatest display of color, because many green algæ can live in water of that degree of heat. In the hottest waters (185° F.) only white filamentous bacteria are found, which gradually become of a sulphur-yellow color at 175° F. This yellow growth is due to a species of *Beggiatoa*, a plant which may be classed with the BACTERIACE.E, and which, during life, deposits sulphur granules.

As the water cools down, other forms of vegetable life appear, give variety to the colorations and give beauty to the borders of the hot pools and overflow channels leading from them. The sequence of temperatures and of colors is somewhat as follows: white, 160° F.–185° F.; yellow, 145° F.–160° F.; red, 130° F; green, 110° F.–130° F.; green-orange-brown, 95° F. There are variations, however, in the sequence of these colors, owing to various environmental conditions. Thus, in the Black Sand Basin and Specimen Lake, the range of color is somewhat this: White, yellow, flesh pink, bright pink, yellowish-green, emerald.

Studying the growths at the several temperatures, we find *Leptothrix laminosa* growing at 135° F.–185° F.; *Phormidium* at 165° F.; *Beggiatoa* at 150° F.–165° F., and *Spirulina* at a lower temperature

Gleocapsa, a blue-green alga, is found growing on the sides of geyser cones, where steam is escaping, forming there a delicate olive-green coloration. A kind of fibrous sinter is formed by the growth of the little alga, *Calothrix gypsophila*, or the young form of *Mastigonema thermale*, the latter olive colored, and forming the sinter of the crater of the Excelsior Geyser.[1] A coarse sinter is due to a bright red species, *Leptothrix*, a finer variety to *Leptothrix* (*Hypheothrix*) *laminosa*, ranging in color from white to flesh pink, yellow and red to green, as the water cools. Besides the above plants, which belong to the BACTERIACEÆ and the CYANOPHYCEÆ, speaking in a general way, we find that several mosses, MUSCI, are active in the formation of sinter on the slopes below Hillside Spring. These springs issue from the rhyolite slopes beneath the cliffs of the Madison Plateau, and the waters, whose temperatures are 184° F.–198° F., contain both silica and lime in solution, which they deposit in their downward flow. This moss has been determined by Prof. Charles R. Barnes, of the University of Wisconsin, to be *Hypnum aduncum, var. grasilescens*, Br. and Sch.

Besides the sinter and travertine formed by algæ, which remove in the case of the carbonated waters, containing calcium bicarbonate, $Ca (HCO_3)_2$, in solution, the gaseous carbon dioxide, thus depositing calcium carbonate, $CaCO_3$, we have stalactites produced by the growth of several algæ, *Gleocapsa violacea*, *Schizothrix calcicola*, *Synechococcus æruginosus* and *Phormidium* (*Leptothrix*) *laminosum*. An interesting account of the formation of these stalactites has been given to us by Miss Josephine Tilden, who visited, recently, the Yellowstone Park.

In the tepid waters of the overflow basins, for example Specimen Lake, which is produced by the water from the Black Sand Pool, we find extensive diatomaceous beds formed by the growth of numerous diatoms. The water of these areas has encroached on the timber, killing the trees, which stand as bare poles from the treacherous marshes. It is known that these plants deposit silica, as a box, test, or frustule, and it is thus by the activity of the protoplasm that the silicious diatomaceous earths are formed. Samples of this material show the presence of *Denticula valida*, which forms the bulk of the material, *Denticula elegans*, *Navicula major*, *N. viridis*, *Epithema*, *Coccouema*, etc.

[1] Weed, *loc. cit.*

It seems likely to me, in studying the vegetation of hot springs, notwithstanding the statements of Prof. Ernst Hæckel, of Jena, in his interesting work, "Systematische Phylogenie der Protisten und Pflanzen," that the early forms of life on this globe were green unicellular algæ, and from these by retrogression and development other forms have sprung, animal life appearing later than plant, it seems to me, I repeat it, that we must look to the hot springs for the most primitive forms of life, because the temperature conditions are such as more nearly to simulate the conditions existing when this world of ours was in a highly heated state, when seismic phenomena were the rule rather than the exception. It would be necessary in order to establish this proposition to investigate comparatively the vegetation of all the hot springs of the globe, before it would be safe to make such a general declaration as to the origin of vegetal life.

The above ecological sketch sufficiently discloses the salient characters of the interesting geysers and hot springs of the Yellowstone Park. In preparing this article, the writer has endeavored to give the results of personal observation on the spot during eight days of August, 1897. He has been materially aided in its preparation by the following papers and books, which give a somewhat more detailed account of the Yellowstone Wonderland:

BIBLIOGRAPHY.

1887-88, WEED. *Ninth Annual Report U. S. Geological Survey*, p. 619.

1895, CHITTENDEN. *The Yellowstone National Park, Historical and Descriptive.*

1897, HAYNES AND GUPTILL. *Guide to Yellowstone Park.*

1897, DAVIS, in *Science*, N. S., I., p. 145 (July 30).

1897, TILDEN, in *Botanical Gazette*, September, 1897.

UNIVERSITY OF PENNSYLVANIA, November 16, 1897.